west, but in relation to one another, they heaven. Whereas there are millions of fixed bodies that were seen by the ancients to background. These are the sun, the moon and the five planets: Mercury, Venus, Mars, Jupiter and Saturn. Of these, the five planets make themselves noticeable against the background of the fixed stars by their irregular movements, their changing speed, their stations, and the change in direction of their observed motion (occasionally seeming to move retrograde).

ANCIENT LUNAR ASTRONOMY

In the night sky, the most obvious sight is the swift motion of the moon, passing each month in front of the belt of fixed stars comprising the zodiac.[3] For many of the peoples of antiquity the familiar monthly passage of the moon through the zodiacal constellations, combined with its waxing and waning phases, was all the astronomy they knew; it was sufficient for them to orient themselves with regard to the passage of time. This specifically lunar astronomy became developed especially in India, in the system of *nakshatras*, comprising 27 or 28 asterisms (constellations in or near the zodiacal belt) through which the moon moves. As the moon takes about 27 ½ days to orbit once around the zodiacal belt, the division into 27 or 28 asterisms describes its motion such that generally one *nakshatra* is traversed by the moon in the course of one day.

The *nakshatras* are referred to in the ancient Vedic literature as a group of 28 asterisms, but in later works of classical Sanskrit literature they are generally described as being 27 in number. In the Vedic texts, when referring unmistakably to astronomical phenomena (such as the moon with reference to *nakshatras*) it is almost always in connection with determining the times propitious for performing sacrifices.[4] In the Hindu literature often reference is made to certain phases of the moon in relation to the *nakshatras*. Most frequently mention is made of the full moon in a particular nakshatra. For instance, it was cold at the time of the year that was signified by "Full Moon in Phalguni." As the *nakshatra* Purva Phalguni belongs to the constellation of Leo, the full moon in sidereal Leo indicated that the sun was opposite in the constellation of Aquarius, marking the middle of winter in the millennium before Christ when the Vedic texts were written, and also in the half millennium after Christ when the classical Sanskrit literature arose. A system of lunar astronomy based on 28 **lunar mansions** (corresponding approximately to the Indian *nakshatras*) was developed also by the Arabs, who likewise followed the cycle of the year through observing the sequence of full moons in relation to the lunar mansions.

ANCIENT PLANETARY ASTRONOMY

At a more advanced level of astronomy than that of simply observing the phases of the moon and its movement through the zodiac, the Babylonians recognized Mercury, Venus, Mars, Jupiter and Saturn—which at first sight somewhat resemble the appearance of the fixed stars—as stars which move. Through their observation of the night sky, the Babylonians were able to arrive at a fairly complete description of the complex movements of Mercury, Venus, Mars, Jupiter and Saturn. The Babylonian astronomy of the planets, as developed in the last few centuries BC, enabled not only the return of a planet to conjunction with a fixed star but also its "cardinal points" to be calculated and predicted in advanced. (The cardinal points of Mars, Jupiter and Saturn are: first visibility in the morning, first station=start of retrograde motion; opposition to the sun; second station=end of retrograde motion; and last visibility in the evening. The cardinal points of Venus and Mercury are: first visibility as morning star; second station=end of retrograde motion; last visibility as morning star; first visibility as evening star; first station=start of retrograde motion; and last visibility as evening star.)[5]

It was only at the tail end of declining Babylonian civilization, during the Seleucid Era (approximately the third, second and first centuries BC), that Babylonian astronomy reached a level of sophistication capable of describing the complex movements of Saturn, Jupiter, Mars, Venus and Mercury in terms of their cardinal points and their orbits of the zodiacal belt. (The first 360° division of the zodiacal belt—the **sidereal zodiac**—was made by the Babylonians; it comprises twelve 30-degree signs defined in relation to the fixed stars, and originated in the sixth or fifth century BC.)[6]

Already long before the Seleucid Era, the Babylonians observed with keen interest the nightly movements of the planets. Thus the Venus tablets of Ammizaduga (ca. 1600 BC), who was king for 21 years during the Old-Babylonian period (Ammizaduga ascended the throne 146 years after Hammurabi), testify to a detailed observation of the heliacal risings and settings of Venus combined with astrological predictions. Examples include: "The harvest of the land will be successful;" "King shall send greetings to king;" "King shall send challenge of war to king;" "Destruction shall be in the land." A typical example, which refers to Venus as evening star between the seventh day of the month of Ulul and the eleventh day of the month of Ayar, followed by her appearance as morning star on the nineteenth of Ayar, reads:

> If on the seventh of Ulul Venus appeared in the west, the harvest of the land will
> be successful; the heart of the land will be happy. Until the eleventh of Ayar she will
> stand in the west and on the twelfth of Ayar she will disappear, and, having remained

absent seven days from the sky, Venus will shine forth in the east on the nineteenth of Ayar; hostilities will be in the land.[7]

The expression "planets," meaning **wandering stars**, applied strictly speaking (in its original sense) to Saturn, Jupiter, Mars, Venus and Mercury, which resemble fixed stars in their appearance, but which "wander" in relation to the fixed stars. However the term "planets"—as used in antiquity—came to embrace in addition the sun and the moon, which also move in relation to the background of the fixed stars of the zodiacal constellations. The sun and the moon—sometimes known as "the lights"—were regarded as regents of the day and the night, respectively.

The sun is to be distinguished from the moon and the five planets in that the latter may be seen at night, whereas generally speaking the sun alone is visible during the day. Moreover, whereas the moon and the five planets may be observed against the background of the fixed stars, the sun is never seen against the stellar background of the zodiac (except on the rare occasion of a total solar eclipse).

ANCIENT SOLAR ASTRONOMY

Nevertheless the stargazers of antiquity—at least, the more sophisticated among them: the savants of Egypt and Babylon—were quite aware of the passage of the sun through the zodiacal constellations. Thus the Babylonians, by way of observing the motion of the moon against the stellar background, were able to infer from it the zodiacal location of the sun. For example, when the full moon was seen in the middle of Taurus, in conjunction with the bright star Aldebaran ("the Bull's eye," marking the center of sidereal Taurus), they knew that the sun was opposite in Scorpio, in conjunction with the gleaming red star Antares ("the heart of the Scorpion," at the center of sidereal Scorpio). Babylonian astronomers had long established that the rising of Aldebaran coincides with the setting of Antares (and **vice versa**), implying that these two zodiacal stars lie diametrically opposite one another in their respective constellations.[8] Or, when Babylonian sages saw the moon at first quarter in conjunction with the radiant star Regulus in Leo ("the heart of the Lion"), they were able to infer that the sun was in conjunction with the Pleiades in the neck of the Bull (as Regulus is exactly 90 degrees advanced from the Pleiades in the zodiac).

Apart from indirect inference of the sun's sidereal location by observing the phases of the moon together with its passage against the background of the zodiacal constellations, an alternative approach applied in antiquity for determining the sun's zodiacal position was that of observation of the heliacal rising of prominent fixed stars. The sight of a fixed star on the eastern horizon at dawn, rising just ahead of the sun, could give a good indication to knowledgeable observers of the location of the sun in a particular zodiacal constellation. This method was adopted especially in Egypt, where the system of decans originated. Just as the *nakshatras* of ancient India comprise 27 or 28 lunar asterisms, so the decans of ancient Egypt are solar asterisms near the zodiacal belt—36 in number—each ruling a 10-day period of the solar year. For the early Egyptians the rulership of a particular decan over a given 10-day period generally coincided with the heliacal rising of the fixed stars comprising that decan. Later the decans became assimilated to the signs of the sidereal zodiac, each decan becoming simply a 10-degree division of a sidereal sign.[9]

Summarizing this introductory outline so far: India gave birth to an essentially lunar astronomy, based on nightly observation of the moon's passage through the 27 or 28 *nakshatras* (lunar asterisms); the Egyptians developed primarily a solar astronomy, in which the passage of the sun was observed each day at dawn in relation to the heliacal rising of fixed stars belonging to the 36 decans (solar asterisms); and the Babylonians developed above all a planetary astronomy for predicting the "wandering movements" of Mercury, Venus, Mars, Jupiter and Saturn along their orbits through the 12 signs of the sidereal zodiac. (Here, possibly, lies the origin of the Babylonian sexagesimal system—based on the number 60—in the enumeration of the 5 planets with respect to each of the 12 signs of the zodiac.) The astronomical theory developed by the Babylonians also included the sun and the moon together with the five planets. It was the most advanced astronomy of all the peoples of antiquity (in the pre-Christian era), in its use of the sexagesimal system and the division of the zodiac into 360 degrees (12 sidereal signs, each 30 degrees long). Thus the Babylonians were able to predict complicated astronomical phenomena such as solar and lunar eclipses, the occurrence of which remained a mystery to other less mathematically-sophisticated peoples.

II. Babylonian Planetary Religion

THE STARS AS THE ABODES OF GODS

Why were the Babylonians so interested in observing and predicting the movements of the planets? What was the fascination exerted by these heavenly bodies?

The Egyptian observation of the sun's passage through the constellations in relation to the heliacal rising of the 36 decans had an obvious practical use, namely the specification of a calendar dividing the year into 36 (approximately) 10-day periods. Similarly, the Indian observation of the phases of the moon with respect to the *nakshatras* also had a calendrical significance in that it served as an approximate time reckoning for the passage of the seasons. In contrast, the Babylonian observation of the irregular movements of the five planets through the 12 signs of the sidereal zodiac was to no apparent practical end. In order to comprehend why the Babylonians took the trouble to observe and develop a planetary astronomy, it is helpful to try to come to an understanding of the kind of consciousness they had.

From a study of cuneiform texts it is apparent that for the Babylonians the starry heavens were peopled with gods—divine beings—who manifested themselves by way of the heavenly phenomena. For instance, the fixed star realm was the abode of the most ancient gods of the Babylonian pantheon: Enlil, Anu, and Ea. According to the ancient Babylonian text mul.APIN, Enlil's domain was that of the more northerly stars; the realm of Ea embraced the more southerly stars; and Anu had dominion over the middle group of stars, between the northern and southern constellations. During the course of the year the sun was seen to traverse the "ways" of Enlil, Anu, and Ea. Thus in summer the sun passed through the way of Enlil, giving rise to heat; in winter it was to be found in the way of Ea, yielding cold; and in spring and autumn the path of the sun crossed through the way of Anu, resulting in wind and storm.[10]

In the Babylonian religion not only the realm of fixed stars but also the planets were seen as the abodes of divine beings. The heavenly beings associated with the planets came to supplant the older generation of gods. Thus the planetary gods became more important than the fixed-star trinity Enlil, Anu, and Ea. By the sixth century BC most important among the planetary gods was Marduk, the divine being associated with the planet Jupiter. Babylon was regarded as the holy city of Marduk, and his star—Jupiter—was called "the great star," "the planet of the king," or "the royal planet."[11]

JUPITER-MARDUK

Each year during the days following the Neomenia—the new moon falling after the vernal equinox—the New Year's festival was celebrated in Babylon. The festival was primarily in honour of Marduk, to whom long prayers were recited, as well as the Babylonian epic of creation. This creation story pays tribute to Marduk's many achievements as the creator of order in the universe. It relates how through his immense power and great courage he overcame the destructive dragon Tiamat and established divine rule in the world. According to the Babylonian creation epic, Marduk instituted order in the place of chaos, peace instead of war, and brought light into the darkness. In his ordering of the cosmos he "set up for the twelve months of the year three stars apiece" thus relating each month to one of the twelve signs of the sidereal zodiac.[12] He preordained the course of human destiny and the history of mankind. But how was Marduk able to act in guiding the fate and the evolution of the human race?

Marduk called to his assistance the other six planetary gods, each associated with one of the planets (including the sun and moon): Shamash ("god of justice")-sun; Sin ("illuminator of the night")-moon; Ninib ("god of strength")-Saturn; Nergal ("god of battle")-Mars; Ishtar ("goddess of love and fertility")-Venus; and Nebo ("god of agriculture and commerce")-Mercury. Together the council of the seven gods, presided over by Marduk, met periodically to decide the fate of all human beings.[13]

In addition, Marduk designated the holy temple of Esagila in Babylon as his earthly temple, where priests in his service could receive communications from him. The New Year's festival in Babylon culminated with a procession in which the priests bore a statue of Marduk from Esagila down to the Euphrates River and then back to the temple again, as a sign of Marduk's triumphant victory each spring over the darkness of winter.

The main characteristics of Babylonian planetary religion may be summarized as follows:

(1) The stars and planets were regarded as the dwelling places of divine beings, who could be approached through worship and rites in the temples dedicated to them, where communications could be received from them.

(2) the celestial beings associated with the heavenly bodies could thus reveal to human beings—to those competent enough to read the signs of the stars—the purposes of the gods concerning the future. In this way astrological omens were accumulated, inscribed on cuneiform tablets, relevant to the fate of the people, especially as represented in the person of the king. Omens bore reference not only to the king's fortune or misfortune in war and to the economic and social condition of the country, but also to the weather and the prospects for the coming harvest. In short, every aspect of life was considered to be determined by the gods, who revealed their intentions through the movements of the sun, moon and five planets in relation to the fixed stars of the sidereal zodiac—this is the essence of the Babylonian planetary religion.

(3) The king was considered to be the mediator between the gods and mankind. His task was to transmit the will of the gods to his subjects; in this he could call upon the advice of the priests.

(4) The earthly world was thought to reflect the cosmic world. The king of Babylon, aided by his council of priests, in principle sought to mirror on Earth the heavenly council of the planetary gods presided over by Marduk, at which the fate of mankind was decided. The Babylonian king thus became the representative of Marduk on Earth, and (ideally) ought to have been capable of entering into personal contact with the great deity.

Above all, the king of Babylon was expected to be wise, for wisdom was Marduk's most outstanding quality, as comes to expression in the following prayer addressed to Marduk ("the lord of wisdom"):

I would extol the lord of wisdom, the god who...
who... the night, who brings the calm day,
Marduk, the lord of wisdom, the god who...
who... the night, who brings the calm day,
whose raging is like an encompassing storm,
but whose breath is sweet as the morning breeze,
whose fury is unrivalled, whose wrath is the storm flood,
whose heart is appeasable, whose mood is conciliatory,
the springs of whose hands, the heavens can't carry,
who, with his soft hand, stays the flood.
Marduk, the springs of whose hands, the heavens can't carry,
who, with his soft hand, stays the flood.[14]

If Marduk, associated with Jupiter, was the lord of wisdom, what were the essential characteristics of the other six gods in the Babylonian planetary pantheon? A consideration of the Babylonian planetary gods reveals the underlying roots of astrology. Greek astrology simply took over the Babylonian planetary religion and adapted it to accommodate the Greek deities, the Olympian gods. In this form it thus became transmitted westward.

MOON-SIN

Although Marduk, as the god of wisdom, came to occupy the central position in the Babylonian pantheon, there is evidence that in earlier times the moon god—Sin—was regarded as the supreme planetary deity. Sin's holy cities were Ur, Tema and Harran. The latter remained a cult center of Babylonian astral religion until the eleventh century—the last moon temple there being destroyed by the Arabs in AD 1033. It was in Harran that rites of worship to the stars were practiced by the Sabians, until they were wiped out through Arab persecution and the city was destroyed.

The earlier high esteem in which the moon god Sin was held is indicated by the appellation "lord of the gods." Sublimely beautiful, the moon for the Babylonians was the symbol of male beauty and strength. Sin was gracious and merciful. His word was just, and he caused human beings to speak the truth. He was regarded as the creator and preserver of life on Earth, just as the moon symbolizes the cycle of life through its waxing and waning phases. Designated as "he who shines," "illuminator of the night," "illuminator of heaven and earth," Sin represented all that is bright, good, and beneficial to mankind.

He was contrasted with his brother Nergal, the god associated with Mars. Sin and Nergal were the sons of Enlil (in some sources the sons of Anu). As Nergal was sometimes referred to as "the sinister twin," it would appear that Sin and Nergal were in fact twins:

> Sin, the king of heaven, and Nergal, the king of hell... Sin represented the heavenly sphere with all that is bright, good, and beneficial to mankind, whereas his twin-brother Nergal stands for the netherworld and all that is dark, harmful and deadly.[15]

NERGAL-MARS

Nergal was regarded as the god of the netherworld, who spread plagues and pestilence, and who incited war. Probably on account of the highly irregular movements of the planet Mars, Nergal was designated as "he who is constantly wandering about." Other appellations applied to Nergal were "the angry fire-god" and "the god of war." Something of Nergal's sinister aims as the instigator of war is revealed in the **Irra legend**, which describes the war god's attempt to overthrow Babylon's patron god Marduk, the protector of the whole of mankind:

> While under the guidance of their patron god, the people on Earth were pious and god-fearing and lived in unity among each other. Irra, however, disliked this state of affairs and planned to change it. As he could not carry out his plan so long as mankind was under the protection of its patron god, he lured the god of Good into the netherworld and seized the reins of power on Earth. As their new ruler, he first perverted the minds of men so that they began to fight among each other; the ensuing war gave him an opportunity to finish his work of destruction and annihilation on Earth. When the patron god returned, he found his cities in ruins and his worshippers slain. (The poem ends with an exhortation to mankind to appease the evil god by allotting a place in their cult to his service, so that he might spare them from another catastrophe like the one described. The subject matter of the legend as well as its treatment implies that, in his quality as a planet, the patron god was unable to protect the community of his worshippers during his periodic absences from the nocturnal sky.)[16]

SUN-SHAMASH

As neither the sun, nor the moon, nor any of the five planets remains all the time visible in the sky, none of the seven planetary gods could be regarded as the sole ruler. For this reason Marduk had to call the council of all the seven gods in order to govern. Although in the course of the history of Babylon Marduk established himself as the president of the council, the moon-god Sin was in earlier times called "lord of the gods." Similarly, during the first dynasty of Babylon—of which Hammurabi was the sixth ruler—the sun-god Shamash was glorified as "king of the gods." Hammurabi in fact gave his son and successor the unusual name "Sa-amsu-iluna," which means "the sun is our god." The name "Hammurabi" itself points to worship of the sun, as "Hammu" was an old Semitic name for the sun god.[17]

Shamash, the sun god, was the son of Sin and the brother of Ishtar. He was revered as the great benefactor of mankind and the Earth, kindling life with his light and warmth. However, he could also burn up crops and dry out rivers. Possibly on this account he did not become considered as leader of the Babylonian pantheon. Nevertheless, by virtue of the sun's radiance—expressive of the sun god's goodness—Shamash was looked to as "the god of justice" and "the personified right." The absolute regularity of the movement of the sun endowed Shamash with the quality of infallibility.

SATURN-NINIB

A mysterious relationship was seen between Shamash and Ninib, the god associated with the planet Saturn. Ninib—also known earlier as "Ninurta"— was designated "the sun of the night," manifesting during the night qualities similar to those that Shamash displayed during the day. Thus Ninib was referred to as "the stable one," "the signifier of equity and justice," and Saturn was termed "the star of righteousness" and "the star of justice and right." But just as Shamash was seen to have a dual nature—not only did he bestow life but he could also destroy it—so Ninib had a negative aspect. Alongside his positive nature as the bearer of justice and righteousness, Ninib was also known as "the god of sickness and death."

MERCURY-NEBO AND VENUS-ISHTAR

In turning to look at the last two gods of the Babylonian pantheon—Nebo (Nabu) and Ishtar: the gods associated with Mercury and Venus—we have to take account of something that is now taken for granted. When we see Mercury and Venus as morning stars—visible on the eastern horizon before sunrise—we have no difficulty in identifying them again as the same planets when they later become visible on the western horizon as evening stars. However, there was a time in the history of mankind when the identity of Mercury as morning star and evening star was not recognized. Similarly, only at a certain point in history did Venus as evening star become identified as the same planet as morning star.

Before the advent of Assyriology, Pythagoras was generally attributed with the discovery that Venus as evening star and as morning star is one and the same planet. Now, however, through the excavation and decipherment of thousands of cuneiform tablets, it is clear that the Babylonians knew of this identity before Pythagoras. But in earlier times—as is evident from the more ancient cuneiform sources, those written in Akkadian—Venus as evening star and Venus as morning star "were conceived as two distinct planets represented in the divine hierarchy by a pair of twin sisters... a period of more advanced astronomical knowledge is reflected in the name 'house of the twins,' which characterizes the

Venus-temple at Nineveh."[18] In the Akkadian texts Venus as evening star was called **Beltan** ("our lady"), and as morning star was referred to as **Dilbat** ("the herald"). It was only later that the single name Inanna (=Ishtar for the Babylonians), the goddess associated with the planet Venus, was employed for Venus both as morning and as evening star, implying that by then the identity had been recognized.

Similarly, a dual designation is found in connection with the planet Mercury in some ancient sources. Such a text is referred to in the following:

> After having assigned their respective numbers to the fixed star deities Anu, Enlil, and Ea, the text proceeds to Sin, Shamash (and his alter ego, Adad), Marduk, Ishtar, Ninurta, and Nergal, and finally names, together in one line, the pair of gods Bilgi and Nusku, who are further characterized as 'the companions of Shamash.' As Sin represents the moon, Shamash the sun, Marduk the planet Jupiter, Ishtar the planet Venus, Ninurta the planet Saturn, and Nergal the planet Mars, it is obvious that the two gods Bilgi and Nusku represent the remaining seventh planet, Mercury...(Thus we can assume that) in the early stages of astronomical observation, Mercury, the morning star, and Mercury, the evening star, were likewise conceived as two different planets represented by two different gods, and that only at a later period their identity was recognized.[19]

The earliest references to the worship of Nusku and Bilgi in Akkadian mythology stem from the second half of the third millennium BC. Later came the discovery of their identity, signified in one text by the statement: "Nusku is Nabu."[20] From then onward the single name Nebo (Nabu) was utilized when referring to the god of the planet Mercury.

Nebo's role in the Babylonian pantheon was to carry Marduk's wisdom to the level of human intelligence. The god of Mercury was therefore seen as the source of human wisdom. He was the son of Marduk, born of Marduk's union with his divine consort Ishtar. At the council of the gods summoned periodically by Marduk, Nebo not only acted as "the divine scribe," but also as "the bearer of the divine tablets of destiny" which were preordained for mankind by the planetary gods at their council meetings. In this capacity Nebo was invoked as "the guide of destiny." More generally he was regarded as "the god of commerce and agriculture." With respect to his beneficial influence upon nature, he was known as "the god who opens fountains and makes the longed-for fruits grow." Owing to the swiftness of Mercury's appearances and disappearances, giving the impression that Nebo was able to behold whatever was going on anywhere, he became designated as "the overseer of the entirety of heaven and earth." He could interpret the divine mind of his father Marduk in the latter's wise leadership of the gods and mankind.

To Nebo, beneficent son, director of the hosts of heaven and earth, holder of the tablets of knowledge, he who hath grasped the writing reed of destinies, lengthener of the days, vivifier of the dead, establisher of light for the men who are perplexed...[21]

Lastly, Ishtar was the goddess associated with the planet Venus. She was the divine consort of Marduk. As the daughter of the moon god, she was "Sin's heroic daughter." On account of the radiant appearance of Venus, she was called "the shining one" and her star was "the most brilliant of the stars." She was titled "the lady of heaven and earth" and Uruk was her city, where the main Ishtar sanctuary of the Babylonians was to be found. Ishtar was invoked as "the goddess of love and fertility." She was the personification of womanhood. As Marduk's divine consort she was fearless in assisting him in the struggle to establish law and order. She could be turned to for assistance in the cause of righteous battle, especially in the struggle to help the downtrodden and oppressed, as emerges in the following invocation (ca. 1500 BC):

I implore you, lady of ladies, goddess of goddesses, Ishtar, queen of all habitations, leading mankind aright... You are strong, you are ruling, your names are excellent. Indeed you are the luminary of heaven and earth, valiant daughter of Sin, who wields the weapons, who orders the battle array... You give decisions for all mankind in justice and equity; you look with favour upon the mistreated and oppressed (and) daily give them true judgement. Have mercy, lady of Holy Eanna, the pure treasury... Valiant Ishtar, your valour is great, shining torch of heaven and earth, brilliance for all human habitations, fierce in the battle that cannot be withstood, brave in the mêlée, fire that blazes against the enemy, who causes the destruction of the fiercest...[22]

III. The Names of the Planets

FROM THE BABYLONIANS TO THE GREEKS

It is not easy for the modern human being to comprehend the Babylonian mode of consciousness. Far from seeing the fixed stars and the planets as the abode of divine beings, they are regarded in modern astronomy simply as material or gaseous bodies emitting or reflecting light and, in the case of the sun and the fixed stars, transmitting vast amounts of energy. Yet in the astrological tradition a conception of the planets and signs of the zodiac as bearers of higher principles of consciousness has been handed down right to the present time. The roots of the astrological tradition are to be found in Babylonian culture, but it is especially the ancient Greeks who inherited the Babylonian tradition and who transformed it into the form in which it became transmitted further.

Greek culture, with its distinctly intellectual heritage, is more easily accessible to a modern understanding than the remote culture of ancient Babylon. This is because the Greeks, beginning with the Greek philosophers, developed the principle of applying rational thought to the quest for knowledge, and science has continued this principle further. In contrast, the Babylonians acquired knowledge using a different kind of consciousness than the rational scientific consciousness prevalent in modern civilization. For instance, it is reported that the Babylonian king could "in his dream vision receive from the gods direct orders which it was his task to transmit to his subjects."[23] The faculty of rational thought, as such, does not appear to have played a significant role with the Babylonians until the last centuries of their civilization. Primarily they received knowledge by way of "dream vision." It was in a dreamlike state of consciousness that they saw the world around them. In this state of consciousness they beheld in dream vision the stars and planets as the abodes of the gods— ensouled beings of a higher order.

The consciousness of the ancient Greeks represents a stage of transition from the dreamlike consciousness of the Babylonians to the rational, scientific consciousness prevalent in modern civilization. The ancient Greeks, although they developed the faculty of rational thinking, also had from the earliest times a pantheon of gods similar to that of the Babylonian pantheon, indicating that they too originally had access to levels of cognition other than the purely rational. Thus the Greeks had no difficulty in assimilating the Babylonian planetary religion. The Greek pantheon—the Olympian gods—simply became substituted for the Babylonian pantheon:

When the Greeks learnt from the Babylonians to distinguish the five wandering stars from those which form the constellations... they consecrated each one to a divinity: Nabu, divine scribe, master of wisdom, as was Thoth in Egypt, became (as did the latter) assimilated to Hermes. Ishtar, the great goddess of love and fertility, naturally became Aphrodite. Ares took the place of the red Nergal, the war god and great purveyor of fires, and Zeus, king of Olympus, that of Marduk, the leader of the Babylonian pantheon. Lastly, Ninib, the malign divinity of combat, was identified with Cronus, the cruel murderer of his father and children. Thus for each of the Babylonian gods was substituted—as the master of planet—a Greek god which shared some resemblance of character with it.[24]

In the transmission of the cosmic religion from Babylon to Greece, initially the Greeks remained true to the principle of seeing the planets as the abodes of divine beings. Mercury was designated as "the star of Hermes," Venus as "the star of Aphrodite," Mars as "the star of Ares," Jupiter as "the star of Zeus" and Saturn as "the star of Cronus." In addition, the moon was spoken of as "the star of Selene" and the sun as "the star of Helios." This form of designation corresponds exactly with that of the Babylonians, for whom Mercury was "the star of Nebo," Venus "the star of Ishtar," and so on. But the more intellectual mode of consciousness that increasingly asserted itself in Greek culture led to a rationalization whereby the planets later became referred to simply by the names of the gods associated with them: Mercury then became called Hermes, Venus-Aphrodite, Mars-Ares, Jupiter-Zeus, Saturn-Cronus, the moon-Selene, and the sun-Helios. The changeover from the one mode of designation to the other possibly occurred as a consequence of the development of the Greek science of horoscopy—primarily in Alexandria—in the second century BC. (And simple convenience may have played a role.) In the case of the most famous Greek astronomer and astrologer, Claudius Ptolemy, who lived in Alexandria in the second century AD, although both modes of designation are to be found in his astrological work the **Tetrabiblos**, more generally Ptolemy writes "the star of Cronus," "the star of Zeus," etc., and only infrequently merely "Cronus," "Zeus," and so on.[25]

It is interesting to note that Ptolemy did not use the traditional Greek names for the planets, which appear to have been derived from astronomical observation. The earliest mention of these names is in a work quoted by Apollonius of Rhodes, who flourished around the middle of the third century BC. According to these Greek astronomical designations, Saturn was called **Phainon**, meaning "the luminous one;" Jupiter was **Phaethon**, "the radiant one;" Mars was **Pyroeis**, "the fiery one;" Venus was **Phosphoros**, "the light-bearer;" and Mercury was **Stilbon**, "the scintillating one."[26] These names are clearly expressive of the appearances of the five planets.

THE ROMANS AND THE EGYPTIANS

The Romans never translated the names Phainon, Phaethon, Pyroeis, and Stilbon into Latin. Only Phosphoros became translated: as **Lucifer** ("the light-bearer"), which was sometimes used by the Romans to designate Venus as morning star. As evening star Venus was occasionally referred to by them as **Vesper**, which is the Latin equivalent of the Greek name **Hesperos** that was often used by the Greeks to indicate Venus as evening star. The terms Phosphoros (morning star) and Hesperos (evening star) were used as early as the eighth century BC by the blind Greek poet Homer.

The Latin designations for the planets (Saturnus, Iuppiter, Mars, Sol, Venus, Mercurius and Luna) stem from the Roman pantheon of gods. These are the names with which we are familiar in their anglicized form (Saturn, Jupiter, Mars, sun, Venus, Mercury and moon). Like the Babylonians and the early Greeks, who designated the planets as the dwelling places of the gods, the Romans too originally referred to each of the planets as "the star of" the god concerned (in this case belonging to the Roman pantheon). For example, Venus was **stella Veneris** ("the star of Venus," i.e., the star of the goddess Venus) and it was only later—around the time of Christ—that the more convenient shortened expression **Venus** was adopted. This difference implies that earlier the planet was seen by the Romans as the abode of the goddess Venus and later the planet itself was beheld as Venus. In this way the planets became substituted for the gods of the Roman pantheon, just as among Greek astrologers the planets came to take the place of the gods of the Greek pantheon.

On the one hand it is apparent that the gradual disappearance of the Greek gods—a process which was already underway around the time of the birth of Christianity—occurred in the wake of the emergence of Greek astrology, which arose in Hellenistic Egypt in the second and first centuries BC. On the other hand, after the Roman conquest of Egypt (30 BC), probably influenced by the astrology flourishing there, a parallel process—that of the gradual disappearance of the Roman gods for whom the planets became substituted—began to take place in Roman culture.

There still remains the question of the names given to the planets by the ancient Egyptians. Here the problem is that although the planets are sometimes depicted on the bas-reliefs of certain Egyptian temples and monuments—one early example being the tomb of Senmut, ca. 1500 BC—scarcely any literary references to the planets have been found. One such reference simply refers to Saturn as "the western star," Jupiter as "the star of the southern heaven," Mars as "the eastern star," Venus as "the god of the morning" and Mercury as "the ever-reappearing nomadic star."[27] In terms of the Egyptian pantheon of gods, there are only a few hieroglyphic inscriptions which relate the planets to the Egyptian gods. One refers to Horus "the bull of heaven" (=Saturn); Jupiter as the star of Osiris; Mars as the star of red Horus; Venus as the star of Isis; and Mercury is not mentioned.[28] Another text has been translated as follows:

> of the five living stars: Horus the bull (=Saturn), it is the star of Re. Horus the red (=Mars), it is the star of the fierce lion. **Sbg** (=Mercury), it is the star of Thoth. The morning star (=Venus), it is Horus, son of Isis. Horus of the secret (=Jupiter), it is the star of Ammon. These are the names of the five living stars together with all gods constituting their names.[29]

Apart from these scanty literary references concerning the Egyptian gods in relation to the planets, the Greek writer Pliny indicates that the gods of the Egyptians were related to the planets as follows: Nemesis to Saturn, Osiris to Jupiter, Heracles to Mars, Isis to Venus, and Apollo to Mercury (*Natural history* II, 34). And according to the Roman author Sextus Empiricus (*Adversus mathematicos* V, 31), the "Egyptians" (probably meaning Nechepso and Petosiris) likened the sun to a king, the moon to his queen, and the five planets to their attendants.

NECHEPSO AND PETOSIRIS

Here we have to take into consideration the fact that for the majority of Greek and Roman authors writing in the post-Christian era on the theme of astrology, the "Egyptians" generally means Nechepso and Petosiris.[30] The names "Nechepso and Petosiris" designate the authors of the hermetic astrological texts comprising the original "astrologers' bible."[31] These writings, only fragments of which remain, are in the form of instructions of the priest Petosiris to the king Nechepso. Firmicus Maternus (fourth century AD) in his great work on astrology, the eight books of the *Mathesis*, speaks of "most powerful Hermes" entrusting his secret to these "divine men,"[32] And a papyrus of AD 138 states that Nechepso and Petosiris "established" their teachings upon those of Hermes.[33]

Nechepso and Petrosiris composed their astrological works in Hellenistic Egypt, probably in Alexandria, during the second century BC. Fragments of these works were collected by Ernst Riess.[34] Some of these fragments derive from a revelation-text in which Nechepso the king, under the guidance of the wise priest Petosiris, is granted knowledge of the truths of horoscopy by way of a vision. In fragment 1 Nechepso recounts a vision in which he beholds the stars and is entranced by the beauty of the starry heavens. And in fragment 33 Petosiris is described as one "who communed with many hosts of gods and angels."[35]

The transformation from the Babylonian planetary religion of the gods to the science of horoscopy (genethlialogy)—although it began in the fifth century BC with the Babylonians themselves—can be traced back to "Nechepso and Petosiris," the names chosen by the authors of the original astrologers' bible, who were hermetic savants living in or near Alexandria in the second century BC. It is with them, through their revelations concerning the truths of horoscopy, that Greek astrology—or at least the rudiments of that which later became practiced as astrology throughout the world—became born. They were the recipients of the Babylonian teachings which became transmitted to Alexandria after Alexander the Great's conquest of Babylon. Moreover, they knew also of the teachings of the Hellenistic philosophers concerning Greek astronomy and physics, including the teaching of the four elements. Their revelations concerning astrology represented a synthesis of Babylonian astronomical methods and celestial omens with Greek astronomy and physics, together with their native Egyptian star lore. From a loose conglomeration of teachings they created a system, thus laying the foundations of the science of astrology. This does not mean that they originated horoscopy—for the earliest horoscopes are of Babylonian origin[36]—but that they systematized what was previously unsystematic.

Through this systematized astrology of the hermetic savants the gods became **planetary principles**. The most striking example of this is the planetary week, the origin of which is attributed to Petosiris.[37] The planetary week is based on the principle that time is under the control of divine beings, each of whom rules in turn. Initially these divine beings were conceived of as the gods of the planets, but later it was the planets themselves which assumed rulership of the seven days of the week. Originally the planetary week started with the day of Cronus (Saturday), as the god Cronus was believed to have ruled over "the golden age," the paradisiacal period at the start of evolution. "Saturn's day" was also held to be the Sabbath, according to Tacitus (first century AD), because "of the seven stars which rule human affairs Saturn has the highest sphere and the chief power" (*Historiae* V, 4). But shortly after the planetary week spread to Rome (around the time of Augustus), the cult of the sun became so widespread that the day of the sun became instituted as the first day of the week—a development that took place quite independently of Christianity. The use of the planetary week—Sunday, Monday, Tuesday (Mars), Wednesday (Mercury), Thursday (Jupiter), Friday (Venus) and Saturday—throughout the whole world is a sign of the triumph of the astrological religion based on the planets, which took the place of the pagan gods. With regard to the English names of the days of the week, the Anglo-Saxons substituted the names of their own deities corresponding to those of the Roman gods, i.e., Tiw (Tyr) instead of Mars, Wotan in the place of Mercurius, Thor instead of Iuppiter, and Freya in the place of Venus, giving rise to Tiw's day (Tuesday), Wotan's day (Wednesday), Thor's day (Thursday), and Freya's day (Friday).

The early Christians took up the planetary week, making Sunday the Sabbath day (the day of Christ's resurrection). According to Justin Martyr, writing in the second century AD.:

> It is on what is called the sun's day that all who abide in the town or the country come together... and we meet on the sun's day because it is the first day on which God formed darkness and mere matter into the world and Jesus Christ, our saviour, rose from the dead. For on the day before Saturn's day they crucified him, and on the day after Saturn's day—which is the sun's day—he appeared to his apostles and disciples and taught them...[*]

There was even an attempt to Christianize the planetary week, as is evident from a Greek astrological text, which may be summarized as follows: Sunday, the day of Christ's righteousness (Jesus Christ as "the sun of righteousness"); Monday, day of the Angels; Tuesday (day of Ares), day of the Forerunner, John the Baptist; Wednesday (day of Hermes), day of the Cross; Thursday (day of Zeus), day of the Apostles; Friday (day of Aphrodite), day of the Mother of God, on account of the Crucifixion of her Son on this day; Saturday (day of Cronus), day of God the Father.[39]

The decline of the pagan religion coincided with the arising of Christianity. Indeed, for the Christian, as the highest God—the Logos of the World—himself incarnated for the salvation of mankind, the lesser (intermediary) gods became redundant. When Christianity became established as the state religion of the Roman empire, the fate of the pagan gods was sealed, and the attempt by Julian the Apostate (emperor from 361 to 363) to reintroduce the worship of the gods failed.

But in its metamorphosed form—as astrology—the pagan religion lived on. Despite hostile opposition on the part of some of the church fathers, notably St. Augustine, astrology survived and later even came to be tolerated by the church.

Astrology is alive today and represents a continuation into our time of the ancient pagan religion, albeit in the guise of planetary principles, which stand in for the pagan gods. It is in coming to an understanding of the pagan gods, therefore, that the roots of astrology can be grasped.

IV. The Pagan Gods

CULTURAL TRANSMISSION

A comparative study of Babylonian, Greek and Roman mythology reveals a high degree of agreement between the gods of their three pantheons—at least, between those gods representing the planets—when their respective attributes and characteristics are compared. In Part II ("Babylonian Planetary Religion") the gods associated by the Babylonians with the seven planets (including the sun and moon) were described briefly. From the outset the Babylonians appear to have regarded the planets as the abodes of divine beings, whereas to begin with neither the Greeks nor the Romans seem to have conceived of their gods as being associated with the planets, except in the case of the sun and the moon.

Already through Pythagoras, who is reported to have been initiated in Babylon by the priest Zaratas in the sixth century BC,[40] something of Babylonian wisdom became transmitted to the Greeks. And if the origins of star-worship and astrology in Greek culture are traced back to Pythagoras, it is apparent that he in turn simply carried over these practices from Babylon and introduced them into Greece, albeit in a modified form. Then, after Alexander the Great's conquest of Babylon in 331 BC, an influx of Babylonian teachings began to penetrate into Greek culture on a large scale. In the process of cultural transmission from Babylon to Greece, the Babylonian planetary gods became assimilated and identified with their counterparts in the Greek pantheon of deities. From this point onwards the Greeks, too, looked up not only to the sun and moon but also to the five planets as the abodes of certain divine beings belonging to their pantheon. The Romans, in turn, were influenced by the Greeks. Thus a similar process of assimilation of the Greek gods on the part of the Romans occurred, just as there had taken place earlier between the Babylonian and Greek cultures. And because by this time the Greek gods had become associated with the planets, so also the corresponding Roman deities became related to the planets.

It is therefore helpful to look at the attributes and characteristics of the pagan gods in order to gain an understanding of the cosmic principles which became ascribed to the planets in astrology.

BABYLONIAN SOURCES

Obviously the Babylonian cosmic religion is the most direct source for an understanding of the nature of the planets in astrology. This is substantiated through a consideration of the order in which the Babylonian planetary gods were written (in the later period): Sin, Shamash, Marduk, Ishtar, Nebo, Ninib, and Nergal corresponding to the sequence: moon, sun, Jupiter, Venus, Mercury, Saturn, and Mars.[41] Sin and his son Shamash, who had ruled in ancient times before Marduk assumed the presidency of the council of gods, were placed first. The order in which the remaining five planetary gods—corresponding to Jupiter, Venus, Mercury, Saturn, and Mars—was written had to do with the natures of these deities. Marduk was foremost, as leader of the pantheon, and his planet, Jupiter, thus became the **great benefic** in astrology. At his side was his divine consort Ishtar, "the lady of heaven and earth," who was also regarded as beneficent, and whose planet, Venus, therefore became the **lesser benefic** in astrology. Nebo, as the offspring of Marduk and Ishtar, came next in the sequence. As the source of human wisdom, which can be used for good or evil, Nebo was neither wholly good (like Marduk) nor wholly evil (like Nergal). In fact, as the divine scribe he served all the gods, including the last two in the sequence, Ninib and Nergal, who were viewed as maleficent in nature. In this sense Nebo was morally neutral. Similarly, in astrology his planet, Mercury, became looked upon as neutral, taking on a positive nature when in conjunction with one of the benefics (Jupiter or Venus) and acquiring a negative nature when in conjunction with one of the malefics (Saturn or Mars). Thus, astrologically speaking, the intelligence indicated by Mercury could be used beneficially (e.g., by teachers) or for evil ends (e.g., by thieves).

At the end of the sequence of the Babylonian gods came Ninib and Nergal, whose planets—Saturn and Mars—came to be regarded as the **malefics** in astrology. Nergal, the god of war, was especially destructive, whose evil doings were recounted in the **Irra legend** mentioned earlier. Nergal was viewed as the chief opponent of Marduk; his erratic and violent behavior was seen to be in direct opposition to Marduk's activity as the upholder of law and order. Ninib, on the other hand, although much feared as the god of sickness and death, was not looked upon as wholly evil. Indeed, as the bearer of equity and justice he was seen to aid Marduk in the latter's divine work, and his planet, Saturn, was referred to as "the star of righteousness." In this respect it could be expected that in astrology Mars should have become the greater malefic and Saturn the lesser malefic, if Babylonian influence alone had prevailed in the composition of the original astrologers' bible written by hermetic savants in Hellenistic Egypt. The fact that this is not the case—for in astrology Saturn became the greater malefic and Mars the lesser malefic—would seem to indicate that for the hermetic savants other considerations assumed an important role alongside the original Babylonian teachings. How is this to be understood?

GREEK MYTHOLOGY
SATURN-CRONUS AND MARS-ARES

Here it is necessary to turn to Greek mythology, especially to those gods of the Greek pantheon who, under Babylonian influence, became associated with the planets. It is the Greek gods Cronus and Ares who became related to Saturn and Mars. Just as with the Babylonians, for whom the war-god Nergal was the most evil deity, so among the Greeks Ares, who was continually inciting war, was generally the most hated of all the gods. A stormy and bloodthirsty troublemaker, Ares did not acknowledge any laws and showed no allegiance to any principles or to any other deity, apart from Aphrodite, the goddess of love, for whom he burned with desire. Always spoiling for trouble, Ares had a raging temper and, in blind frenzy, was capable of starting a fight simply for the sake of it, regardless of any moral considerations involved. His fiery and demonic nature corresponds exactly with that of his Babylonian counterpart, Nergal.

On the other hand Cronus, corresponding to the Babylonian Ninib, was generally regarded in a more positive light. Often pictured as a majestic old man holding a sickle in his hand, Cronus was revered by the Greeks as Father Time, the god of time. (The word **chronology**, the science of measuring time, is clearly derived from the same root as the word Cronus.) As the lord of time, Cronus also ruled over death. Knowing the appointed time when a human being's earthly sojourn would come to an end, Cronus would then approach with his sickle in order to reap the harvest of death, namely the immortal soul of the human being, which became harvested by Father Time when the mortal body was laid to rest. Cronus was also looked upon as a promoter of morality and of civilized life.

Moreover, with his scythe he became venerated as a god of agriculture. In some sources Cronus was spoken of as king of the Golden Age, ruler over mankind in a remote past, during a paradisiacal period at the beginning of human history. In this connection he was thought of as father of the gods, earlier the supreme world ruler. But what was the negative side of Cronus' nature?

It is Hesiod's work **Theogony**, the genealogy of the gods, which was the source for the Greeks of the conception of Cronus as a malign deity. In this work Cronus is described as the youngest son of Ouranos ("Heaven") and Gaia ("Earth"). Together with the other offspring of this marriage, known as the Titans, Cronus challenged the rulership of Ouranos. In the ensuing struggle, which the Titans won, Cronus castrated his father. Although he was the youngest of the Titans, Cronus was looked upon as their leader, and so he became established as the supreme world ruler. He married his sister Rhea, who bore him six children: Hestia, Demeter, Hera, Hades, Poseidon, and Zeus. But such was the jealous determination of Cronus to retain the reins of power that he murdered each newborn child by eating it alive, so that none of his children should ever succeed him to the throne. After having murdered the first five offspring in this way, Rhea—prior to the birth of Zeus—conceived of a plan whereby her sixth child could be saved. She managed to hide Zeus at birth, and to enable him to grow up in safety. Later, when Zeus came out of hiding, he was able to overthrow Cronus and assume the rulership of the universe. Zeus then introduced a new world order, creating all things anew by surrounding them with "ineffable ether," and became the highest of the gods, reigning from Mt. Olympus. Cronus was banished, downcast and dejected, to lead the lonely life of a fallen and forgotten god.

From the description of Cronus presented by Hesiod, a gloomy picture emerges of a sullen and malign divine being. Having castrated his father and murdered his children, Cronus appeared to many Greeks as a terrible, evil deity. It is perhaps this aspect of Cronus which made the god seem to be even more evil than Ares in the eyes of the hermetic savants who wrote the original astrologers' bible. Despite the positive side of Cronus, which was otherwise generally acknowledged by the Greeks, it could be that the awfulness of Cronus' deeds in the history of the gods, as depicted by Hesiod, was such as to make him appear to the hermetic astrologers more evil even than the war-god Ares. Possibly for this reason, then, the planet Saturn became in astrology the great malefic and Mars the lesser malefic, thus reversing the Babylonian order of malignity. The movement of the planet Saturn, the slowest planet, orbiting in the highest and coldest regions of heaven, was seen by the hermetic astrologers as expressive of the withdrawn, downcast, sullen, lonely and forgotten god, Cronus. Similarly, the red-glowing planet Mars, with its highly irregular forward and retrograde motion, was a clear manifestation of the fiery, malignant, unpredictable Ares.

JUPITER-ZEUS, VENUS-APHRODITE, AND MERCURY-HERMES

It is evident from the foregoing that there is a fair degree of correspondence between the natures of the Babylonian gods Ninib/Nergal and those of their Greek counterparts Cronus/Ares. Similarly, there exists a fairly close correspondence between the remaining planetary gods of the two pantheons. Thus the attributes of the Babylonian deity Marduk agree well with those of the Greek god Zeus, whose planet came to be regarded as Jupiter. Just as Marduk was revered as the president of the council of Babylonian gods, so Zeus was considered the leader of the community of gods established on Mt. Olympus, the tallest mountain in Greece. Zeus was "the god of heaven," referred to with the title "Father," i.e., lord and master, protector and ruler. He was strongest, mightiest, wisest, and he was righteous, omnipotent, mysterious and enigmatic. He ordered the cycle of the day, the seasons of the year, the passage of the ages. He foreordained human destiny, and he was the guardian of law and morals; he granted kings their authority and guarded their power and rights. He was also the protector of house and family, of the weak and helpless, and of all who turned to him for help. He determined all that took place and guided everything toward its aim or conclusion. He manifested himself in the weather, in wind and rain, and in the atmosphere in general, but most especially in thunder and lightning, which earned him the titles "Lightener" and "Thunderer."[42] The majestic radiance of Jupiter and the planet's steady motion were seen as manifestations of kingly Zeus, the lawgiver.

With respect to Aphrodite, the Greek goddess whose planet became identified as Venus, it is only in her capacity as the goddess of love and beauty that she corresponds to the Babylonian Ishtar. For the Babylonians, Ishtar was the consort of Marduk; for the Greeks, Aphrodite was married to Ares. Ishtar was "Sin's valiant daughter" to whom the Babylonians turned in the cause of righteous war, whereas the Greeks invoked the goddess Athena in going to battle for a just cause rather than Aphrodite, who was looked upon as the embodiment of everything feminine and was decidedly unwarlike in nature. Nevertheless, in her title as "the queen of heaven" Aphrodite certainly would seem to correspond to Ishtar, "the lady of heaven and earth." For the Greeks Aphrodite was above all the goddess of beauty, and the brilliant radiance of her planet, Venus, was regarded as a manifestation of this divine beauty.

A clear correspondence is evident between the Babylonian Nebo and the Greek Hermes, whose planet was recognized to be Mercury. The youngest of the Olympian gods, Hermes was the messenger and servant of Zeus and the other gods. He was the leader of souls to the Underworld, the guide of travelers, the patron of writing, literature, languages, and the art of speaking, as well as of the sciences, especially mathematics and astronomy. He was the inventor of the lyre and the flute. Above all he was the god of agriculture and commerce (the latter also in a negative sense as the patron of thieves). Hermes was renowned for his most powerful word ("logos"), which earned him the title **Hermes logios** ("eloquent Hermes"). He was depicted with winged feet and in swift motion, which came to manifestation in the rapid movements of his planet Mercury.

MOON: SELENE-ARTEMIS-HECATE

With regard to the sun and moon, associated with the gods Shamash and Sin in the Babylonian pantheon, the Greeks saw more than one deity in connection with these heavenly bodies. Moreover, a fundamental difference in conception emerges in that for the Babylonians the moon-god Sin was male, whereas the Greeks associated only goddesses with the moon, namely: Selene, Artemis, and Hecate. These Greek moon-goddesses represent three aspects of the moon. Selene was most closely associated with the actual appearance of the moon itself, when the moon was visible in the heavens, high above the Earth. Artemis was linked especially to the rising and setting moon, when the moon was on the horizon. And Hecate was the goddess of the moon in the depths, when the moon was invisible beneath the horizon, hidden below the Earth.

Selene was the personification of the moon itself. Like Aphrodite, Selene was also a goddess of beauty, seen manifesting herself on moonlit nights in the beautiful radiance of the moon. Owing to the close approximation between the woman's monthly cycle and the phases of the moon, Selene was considered to be the patron goddess of menstruation and the physical life of women in general. And as the moon's sidereal rhythm was linked with the embryonic period, Selene was also revered as the goddess of conception and birth.[43] She was thought to help bring about the growth of plants, and to play an important role in matters of health and sickness. Selene was looked upon as the sister and bride of the sun-god Helios.

The moon-goddess Artemis was the twin sister of Apollo, a god who was closely associated with the sun. (See below concerning the relationship between Helios and Apollo.) Whereas Selene was connected with the physical aspect of the moon, Artemis represented more the soul of the moon, which manifested itself especially in the appearances and disappearances of the moon on the horizon. Artemis was elusive and mysterious, of great beauty, eternally youthful, a virgin goddess. Like Selene she was thought to exert a beneficial influence upon the plant world, fructifying and promoting the growth of crops. But over and beyond this she was considered to be mistress of the animal world, guarding herds of cattle, flocks of sheep, etc. Among wild animals she had a special relationship with deer. Indeed, the elusive nature of the deer, fleetingly appearing and disappearing, corresponded closely with Artemis' own character. Consequently she was invoked as the goddess of hunting, and was depicted with a bow and arrow. In addition she was the guardian of children, and she bestowed her blessings at weddings. She was able to come swiftly to the aid of those in need, and could be turned to for advice in all matters to do with healing. She was invoked by travelers, being especially the protector of roads and pathways as well as of ships crossing the sea.

Hecate was thought to be the mistress of the Underworld. As such, according to popular belief, she was of a demonic nature, the goddess of spirits and ghosts. She held the keys to the gates of the Underworld, and it lay in her power to open and close them. She was thus the ruler over souls who came into the Underworld after death. If she took possession of the souls of the living, she could cause madness or even epilepsy. And nightmares were brought about by her approach toward the living during sleep. She was depicted surrounded by dogs—the dog being sacred to Hecate. Above all she was the goddess of magic. Later the Neoplatonists developed a much more elevated conception of Hecate than that of popular belief. For the Neoplatonists she was the Cosmic Soul, the source of all virtue. As mistress of the moon they regarded her as the ruler of the visible world, and she was called "the self-manifesting image of Nature."[44]

SUN: APOLLO-HELIOS

Helios was the sun god of the Greeks, and Apollo was closely associated with the sun. Whereas Helios was the personification of the sun itself, identified more with the sun's concrete appearance, Apollo was the soul of the sun, embodying the sun in an ethical and spiritual sense. There is thus a similar parallel in the relationship between Helios and Apollo as that between Selene and Artemis.

For the Greeks Helios was the symbol of life, i.e., humanity's life on Earth, in the sense that the human being lives "through the days of life which God gives him under the sun" (*Ecclesiastes* 8:15). Helios was seen as the god who drove the chariot of the sun across the sky from east to west each day. Thereby he provided the Earth with light and warmth, kindling life in the whole of nature; and at the same time he ordered the hours of the day and the seasons of the year. With his rays he oversaw land and sea, and could see all that takes place on the Earth. He was especially the god of seeing, and was invoked to heal blindness. Further, he was the protector of holy trees. Like the Babylonian sun-god Shamash, it was also recognized by the Greeks that Helios was capable not only of sending down beneficial warmth but also destructive heat, causing draught and burning crops in times of a heat wave.

Helios and Apollo were sometimes placed side by side. Apollo was a god of radiant light, who enjoyed eternal youthfulness. As such he was the protector of youth, and was seen as the ideal to be aspired to by young males. Through his radiance he was the purifier of the soul, and also of the body from sickness, and he was the forgiver of sins. He was a healing god, and was protector of the soul in the life after death. He helped to establish order in the sequence of the seasons and the months of the year, and promoted farming and guarded flocks. He was the god of medicine, archery, gymnastics, poetry, music and dance. He was the guardian of oracles and prophecy, and he promoted colonization and the founding of cities.

SOLAR LOGOS: CHRIST

The question arises as to whether the Greeks recognized a third level of the sun over and above Helios and Apollo—a level corresponding to the spirit of the sun? This was the case with some Greek philosophers, notably Plato, who spoke of the highest God simply as "the Good." For Plato the Good was the spiritual archetype of the sun; the sun itself was "the Son of the Good" (*Republic* 50b). The Good was therefore "the transmundane sun," whose image was the physical sun.[45] This transmundane sun was attributed with the qualities of the Logos, the creator of the world through the power of the word. Thus in his *Oration to the sun*, Julian the Apostate alludes to the highest God ("the Good") with the following words:

> This divine and wholly beautiful universe, from the highest vault of heaven to the lowest limit of earth, is held together by the continuous providence of God, has existed from eternity ungenerated, is imperishable for all time to come, and is guarded immediately by nothing else than the Fifth Substance whose culmination is the beams of the Sun; and in the second and higher degree, so to speak, by the intelligible world; but in a still loftier sense it is guarded by the King of the whole universe, who is the centre of all things that exist. He, therefore, whether it is right to call him the Supra-Intelligible, or the Idea of Being, and by Being I mean the whole intelligible region, or the One, since the One seems somehow to be prior to all the rest, or, to use Plato's name for him, the Good; at any rate this uncompounded cause of the whole reveals to all existence beauty, and perfection, and oneness, and irresistible power; and in virtue of the primal creative substance that abides in it, produced, as middle among the middle and intellectual, creative causes, Helios the most mighty god, proceeding from itself and in all things like unto itself. Even so the divine Plato believed, when he writes, "Therefore (said I) when I spoke of this, understand that I meant the offspring of the Good which the Good begat in its own likeness, and that what the Good is in relation to pure reason and its objects in the intelligible world, such is the Sun in the visible world in relation to sight and its objects."[46]

For Plato, and also for Julian the Apostate, the sun was begotten in the image and likeness of the Good, the Logos. In the prologue to the Gospel of St. John, the Logos, the divine word, is referred to: "In the beginning was the Word (Logos), and the Word was with God, and the Word was God... and the Word became flesh and dwelt among us, full of grace and truth" (*John* 1: 1-2, 14). According to St. John, who was the most philosophical of the evangelists, the Logos, spoken of by the Greek philosophers, incarnated as Christ. The identification of the Logos (and therewith Christ) with the spirit of the sun was self-evident to early Christians. For example, Anastasios Sinaita wrote: "Helios has the first place in the sky and leads the heavenly dance. So is Christ, who is the spiritual Sun, placed above all the heavenly hierarchies and powers, for He is the leader, the door to the Father" (*Hexaemeron* 5). And St. Gregory of Nazianzus referred directly to Christ in the words of Plato as "God in the kingdom of spirits as the Sun is in the realm of the senses" (*Oratio* 40, 5). It is only against this background of the identification of Christ as the spirit of the sun that it is possible to understand why it was quite natural for the early Christians to place the birth of Christ on December 25, which was at that time reckoned to be the day of the winter solstice, celebrated as the festival of the Invincible Sun. Just as the sun, Helios, was seen to be born (or reborn) on this day—after which each year his light and warmth began to increase—it followed for the early Christians, as they had no historical testimony concerning the date of Christ's birth, that it must have taken place also on this day. For, to them the parallel was clear: the birthday of the physical sun, Helios, must fall on the same day—as a reflection—as the birthday of the spiritual sun, Christ. Thus, by AD 354, December 25, which previously was the festival of the Invincible Sun, began to be celebrated as the day of the Nativity:

> But they name (this day, December 25) also "the birthday of the Invincible sun." Truly, who is as invincible as Our Lord, who cast down and triumphed over death? And if they call this day "the birthday of the sun:" He is the sun of righteousness, of whom the prophet Malachi said: to the God-fearing His name will arise as the sun of righteousness, and salvation is beneath His wings.[47]

THE SPREAD OF ASTROLOGY TO INDIA AND ROME

From its starting point with the hermetic savants of Egypt, based primarily in Alexandria, the Greek astrological tradition spread, becoming transmitted as far as India.

> In the first half of the second century AD there was written in Egypt, and probably at Alexandria, a Greek handbook of astrology on a rather popular level which exercised an enormous influence on vast multitudes of men, though even its title is not now known. For, in AD 149/150 it was translated into Sanskrit in Western India... (and) is the root of Indian astrology.[48]

This work, known as the *Yavanajataka* ("the horoscopy of the Greeks"), which was translated from the Greek by Yavanesvara ("the lord of the Greeks"), signified the introduction of genethlialogy (horoscopic astrology) into India.[49]

Already some two centuries prior to the transmission of genethlialogy from Alexandria to India, with the assimilation of Egypt to the Roman Empire in 30 BC, the way was paved for the spread of astrology to Rome. The earliest extensive astrological work written in Latin is the *Astronomica* by Manilius, who lived during the reigns of Augustus and Tiberius, and was therefore a contemporary of Jesus Christ. From this work (the same applies to other Latin astrological texts) it is apparent that Manilius simply took over the ideas of Greek astrology transmitted from Alexandria and adapted them in terms of Roman mythological conceptions. This was made all the easier since the gods of the Greeks had become assimilated to the Roman pantheon, becoming more or less identified with their Roman counterparts. Zeus was identified with Iuppiter, the leader of the Roman pantheon; Cronus with Saturnus; Ares with Mars; Helios-Apollo with Sol; Aphrodite with Venus; Hermes with Mercurius; and Selene-Artemis with Luna-Diana.

There is, in fact, a remarkable concordance between the attributes of the Roman gods and those of their Greek counterparts, with two exceptions, however, where there was originally something of a divergence. Thus, in the case of Saturnus, whose cult was the oldest in Rome, he was originally regarded by the Romans as a god who bestowed blessings and good luck, being especially beneficent with regard to agriculture. Early on, with the transmission of Greek culture to Rome, Saturnus became identified with Cronus, in the latter's role as king of the Golden Age, who had once led the world in righteousness. With this identification, though, subsequently the negative side of Cronus, as depicted by Hesiod in his work *Theogony*, also became attributed to Saturnus. In the course of time, then, Saturnus acquired a double character for the Romans, just as Cronus had for the Greeks.

Secondly, in the case of Venus, she was originally viewed by the Romans as the protector of gardens. It was only when she became identified with Aphrodite that she became transformed from a garden goddess to a love goddess. Initially she was for the Romans a goddess of all that is beautiful and blossoming in nature, and in human cultural life she bestowed joyfulness, beauty, and charm. She was especially the goddess of fruitfulness and growth of vegetation, exerting a beneficial and healing influence. Through her identification with Aphrodite, Venus then became for the Romans first and foremost the goddess of love. Thus Manilius wrote of the abode of Venus "among the stars, placing in the very face of heaven, as it were, her beauteous features, wherewith she rules the affairs of men. To the abode is fittingly given the power to govern wedlock, the bridal chamber, and the marriage torch."[50] Here it is evident that under the influence of the Greek conception of Aphrodite, Venus had become also for the Romans the goddess of love by the time of Manilius. Similarly, the Greek idea of Cronus is evident when Manilius wrote of Saturnus "cast down himself in ages past from empire in the skies and the throne of heaven, he wields as a father power over the fortunes of fathers and the plight of the old."[51] It is therefore possible to conclude that Greek mythology in its later phase (after it had assimilated the Babylonian planetary religion) was highly influential with regard to the teachings of astrology concerning the attributes of the seven planets.

V. The Planetary Spheres

THE ORIGINS OF HOROSCOPY

There still remains the question as to the origin of the fundamental doctrine underlying genethlialogy. According to this doctrine, the positions of the seven planets against the background of the twelve signs of the zodiac at the moment of the human being's birth are symbolic of one's fate. This doctrine provided the original basis for the casting of horoscopes. But where does it stem from?

As the Babylonians were the first to cast horoscopes—the world's oldest horoscope, from a cuneiform tablet, has been dated to the year 410 BC—it is with the Babylonians that the step from omen astrology to genethlialogy is to be sought. (**Omen astrology** is the name given to the Babylonian practice of reading celestial omens as portents of the future concerning the king, the nation, and the affairs of the land.) In omen astrology the ongoing movements of the seven planets against the background of the zodiacal constellations were observed and were considered to be of significance for the king and the people as a whole. The practice of omen astrology dates back to at least the first dynasty of Babylon (eighteenth to fifteenth centuries BC). It was only at the tail end of the Babylonian civilization that the method of utilizing celestial omens became applied to the individual—with respect to the planetary positions at the moment of birth—in order to predict things concerning the course of the individual's life. This was the birth moment of genethlialogy, around the fifth, fourth, and third centuries BC, during which time the world's first horoscopes were cast by the Babylonians.[52]

The next decisive step came with the hermetic savants of Hellenistic Egypt, who systematized the hitherto fragmentary star wisdom of the Babylonians and added to it. To the original Babylonian horoscopic astrology (concerned solely with planets in signs of the zodiac at the individual's birth) they added their own teachings, and systematically developed genethlialogy. Their teachings, which were completely unknown to the Babylonians, were concerned especially with the significance of the Ascendant and the positions of the planets in the twelve houses of the horoscope at the moment of birth.[53] The hermetic savants of Alexandria thus brought about the birth of Greek horoscopic astrology, which then began its triumphant march across the world, spreading to the Roman empire, and eventually to India. What was the compelling power of genethlialogy? Why was the fundamental doctrine of horoscopic astrology recognized almost universally to be true?

CHALDEAN THEOLOGY

Some insight concerning the basic doctrine of genethlialogy is provided by a study of late Babylonian astral theology, knowledge of which became transmitted to Rome by Julian the Chaldean and his son, Julian the Theurgist, during the reign of the Emperor Marcus Aurelius (AD 121-180). They were the authors of the *Chaldean Oracles*, the teachings of which later became taken up by the Neoplatonists, especially Porphyry and Iamblichus (third and fourth centuries AD). The doctrines conveyed by the *Chaldean Oracles* comprise an esoteric theology which, among other things, serves to explain the basis of horoscopic astrology. (The name "Chaldean" referred originally to the priestly caste of Babylon, but later became attached to their spiritual disciples in Greece and Egypt who acquired an esoteric knowledge of the science of the stars.) According to the Chaldean teaching, the gods of the planets each preside over a sphere, whereby the planetary spheres were conceived of as a nested sequence of concentric spheres extending up to the zodiacal sphere of fixed stars, the Earth being fixed at the center. The first sphere surrounding the Earth, that of the moon, was the sphere of the moon-goddess Hecate. Then came the sphere of Hermes, whose planet Mercury was believed to delineate a sphere on its orbit around that of the moon. The sphere of Aphrodite came next, which was traced out by the planet Venus, whose orbit was thought to embrace that of Mercury. And then came the sphere of Apollo—that of the sun—containing the spheres of the moon, Mercury and Venus within it. Above Apollo's sun-sphere were the spheres of Ares (Mars), Zeus (Jupiter), and Cronus (Saturn). Finally, the seven planetary spheres—moon, Mercury, Venus, sun, Mars, Jupiter, Saturn—were embraced by an eighth sphere, the sphere of the fixed stars comprising the zodiac.

The cosmological teaching concerning the planetary spheres as a nested sequence of concentric shells was known to the Greek astronomer and astrologer Claudius Ptolemy (second century AD). As precisely this cosmological teaching formed the basis of Ptolemy's geocentric astronomical system, it is often referred to as "the Ptolemaic system." For Ptolemy the planetary spheres were "etheric shells" carrying the planets around in their orbits. In the esoteric theology of the Chaldeans the spheres were worlds presided over by the planetary gods—worlds through which human beings pass on their way into incarnation, and which they indwell in the life after death, just as during life itself they dwell upon the Earth, in the sphere of the earthly world.

According to the teachings of the Chaldeans, the human soul enters into the earthly world through conception and birth, thereby acquiring a physical body. Prior to this the soul lives in the heavenly world. On its way into incarnation the soul descends through the planetary spheres, receiving certain characteristics and faculties from each sphere. In so doing it becomes clothed with an astral body. "According to Chaldean doctrine, the fiery spark of the human soul, during its descent from its supercelestial place of origin into the earthly body, acquires a vesture formed out of the substances of the spheres which it traverses."[54] The nature of the astral body, comprising qualities received from the various planetary spheres on the descent into incarnation, is reflected in the planetary configuration at the moment of the individual's birth. This teaching underlies the practice of genethlialogy as originally conceived of by the Babylonian priestly caste. It explains why the Babylonians began to cast horoscopes for the moment of the individual's birth. In the horoscope they had a "map" of the astral body, which, if rightly understood, could reveal the nature of the human being's soul. It could also express the meaning of one's destiny, by providing a glimpse into the journey through the planetary spheres on the way into incarnation upon the Earth—the human being's life on Earth being the goal and purpose of the journey into incarnation. At the end of earthly life, according to the Chaldeans, the human being then returns to the celestial world, journeying back through the planetary spheres in the reverse direction.

> The notion that the soul descending from heaven takes on the characteristics of the planetary spheres through which it passes, before it enters into corporeal existence, and that after death it makes its journey through the heavens in the reverse direction and with opposite effect—this derives from the same religious circles as those in which the doctrine of the voyage of the soul through the spheres had developed: the later Babylonian astral theology.[55]

As mentioned above, the Chaldean teachings became transmitted to the Neoplatonists; indeed, it is thanks to the Neoplatonist Proclus (fourth century AD) that the texts of the Chaldean Oracles became preserved. Another important source for Neoplatonic thought is Macrobius' *Commentary on the Dream of Scipio*, in which the soul's descent into incarnation through the planetary spheres is described. In this work the qualities acquired by the soul in each sphere are listed as follows: "reason and understanding" in the sphere of Saturn;

"the power to act" in the sphere of Jupiter; "a bold spirit" in the sphere of Mars; "sense-perception and imagination" in the sphere of the sun; "the impulse of passion" in the sphere of Venus; "the ability to speak and interpret" in the sphere of Mercury; and "the function of molding and increasing bodies" in the sphere of the moon.[56] These qualities were bestowed upon the incarnating soul by the gods of the planetary spheres. Thus Ares (Mars sphere) endowed the human being with "a bold spirit," Aphrodite with "the impulse of passion," Hermes with "the ability speak and interpret," etc. Here the inner relationship of horoscopy to the planetary religion of the gods is revealed.

Although Christianity suppressed the pagan religion, it did not reject the cosmological teaching of the planetary spheres. In fact, half of the Chaldean/Neoplatonic doctrine concerning the planetary spheres became taken up—in metamorphosed form—to become an intrinsic part of Christian eschatological teaching. The half of the Chaldean/Neoplatonic doctrine that became Christianized is the nonastrological half, i.e., the descent of the soul into incarnation became left out, and only the ascent of the soul through the planetary spheres in the life after death became taken up. Instead of the pagan gods presiding over the planetary spheres, the Christian teaching substituted in their place the spiritual hierarchies according to the doctrine attributed to Dionysius the Areopagite, who was a pupil of St. Paul. The spiritual hierarchies were conceived of in ascending ranks, each allotted to a planetary sphere: Angels—moon sphere; Archangels—Mercury sphere; Principalities—Venus sphere; Powers—sun sphere; Virtues—Mars sphere; Dominions—Jupiter sphere; Thrones—Saturn sphere; Cherubim—fixed star/zodiacal sphere; Seraphim—the crystal heaven. And the human being was envisaged in the life after death ascending through the planetary spheres toward the Empyreum, the realm of the Holy Trinity beyond the planetary spheres and even beyond the fixed stars and the crystal heaven. Dante's great poetic work *The Divine Comedy* depicts this ascent of the soul in the life after death through the planetary spheres and the corresponding ranks of the spiritual hierarchies.

On a closing note, to conclude this study of the history of the planets, it is evident from the foregoing that a reconciliation between astrology and Christianity would have been quite conceivable. For, Christianity always acknowledged the postmortem life of the soul, ascending through the planetary spheres in the life after death. Similarly, ancient astrology acknowledged the preexistence of the soul and its descent through the planetary spheres into incarnation. It would have been but a short step to have linked up the two teachings to form a comprehensive whole. But apparently this was not undertaken until the twentieth century, during which a new era of astrology has begun.

Notes and References

1. Olga von Ungern-Sternberg, "Die Geschichte des goldenen Vlieses", *Hermetika*, vol. 14, 1986, p. 39.
2. Robert Powell, *The Zodiac, A Historical Survey* (San Diego, CA: ACS Publications, 1984).
3. Robert Powell and Peter Treadgold, *The Sidereal Zodiac* (Tempe, AZ: AFA Publications, 1985).
4. P.V. Kane, *History of Dharmasastra*, vol. 5, part I (Poona: Randishi, 1958), pp. 496ff.
5. B.L. van der Waerden, *Science Awakening: The Birth of Astronomy* (Leiden, Netherlands: Noordhoff, 1974), vol. 2, p. 109.
6. Powell, *The Zodiac*.
7. S. Langdon and J.K. Fotheringham, *The Venus Tablets of Ammizaduga* (Oxford, England: Oxford University Press, 1928), p. 11. The most recent edition of the Venus Tablets of Ammizaduga is by Erica Reiner and David Pingree, *Babylonian Planetary Omens* (Malibu, CA: Bibliotheca Mesopotamica, 1975), part I.
8. Powell, *The Zodiac*.
9. Otto Neugebauer and Richard Parker, *Egyptian Astronomical Texts* (3 vols., Providence, RI: Brown University Press, 1960, 1964, 1969), vol. 3, p. 168.
10. Ernst Weidner, "Der Tierkreis und die Wege am Himmel," *Archiv für Orientforschung*, vol. 7, 1931-32, pp. 170-178.
11. For much of the material presented in this study on Babylonian planetary religion, I am indebted to the research of Julius and Hildegard Lewy, see footnotes 15-18, 23.
12. E.A. Wallis Budge, The Babylonian Legends of *the Creation and the Fight between Bel and the Dragon* (London: London University Press, 1921), p. 24.
13. Charles Victor McLean, *Babylonian Astrology and its Relation to the Old Testament* (Toronto: Burlington, 1929), p. 10.
14. Erle Leichty, "Two New Fragments of Ludlul Bel Nemeqi," *Orientalia*, New Series, vol. 28, 1959, pp. 361-363.
15. Hildegard Lewy, "Miscellanea Nuziana II," *Orientalia*, New Series, vol. 28, 1959, pp. 113-129.
16. Hildegard Lewy, "Points of Comparison between Zoroastrianism and the Moon-Cult of Harran," *A Locust's Leg. Studies in Honour of S.H. Taqizadeh* (ed. W.B. Henning and E. Yarshater, London: Lund-Humphries, 1962), pp. 139-161, esp. pp. 148-149.
17. Julius Lewy, "The Old West Semitic Sun-God Hammu," *Hebrew Union College Annual*, vol. 18, 1944, pp. 429-481.
18. Hildegard and Julius Lewy, "The God Nusku," *Orientalia*, New Series, vol. 17, 1946, pp. 148-149.
19. Ibid., pp. 147-149.

20. Ibid., p. 155.

21. E.A. Wallis Budge, *The Babylonian Story of the Deluge and the Epic of Gilgamish* (London: London University Press, 1920), p. 20.

22. E. Reiner and H.G. Güterbock, "The Great Prayer to Ishtar," *Journal of Cuneiform Studies*, vol. 21, 1967, p. 255.

23. Hildegard Lewy, "The Babylonian Background of the Kay Kaus Legend," Archiv Orientalni vol. 17, 1949, pp. 28-109.

24. Franz Cumont, "Les noms des planètes et l'astrolatrie chez les Grecs," *L'Antiquité Classique*, vol 4, 1935, p. 7.

25. Ptolemy, *Tetrabiblos* I, 4 (trsl. F.E. Robbins, Cambridge, MA and London: Loeb Classical Library, 1940, p. 35, footnote 3).

26. Cf. O. Neugebauer and H.B. van Hoesen, *Greek Horoscopes* (Philadelphia: American Philosophical Society, 1959), p. 1.

27. W. and H.G. Gundel, "Planeten," *Pauly Real-Encyclopädie der classischen Altertumswissenschaft*, vol. 20, part II, cols. 2017-2184, esp. col. 2027.

28. Ibid.

29. Otto Neugebauer, "Demotic Horoscopes," Journal of the American Oriental Society, vol. 63, 1943, p. 121.

30. David Pingree, "Petosiris," *Dictionary of Scientific Biography*, vol. 10, pp. 547-549.

31. Cf. Wilhelm and Hans Georg Gundel, *Astrologumena* (Wiesbaden, Germany: Sudhoffs Archives, 1966), p. 14.

32. Firmicus Maternus, *Matheseos* III, i, 1 (trsl. J.R. Bram, *Ancient Astrology, Theory and Practice*, New Jersey: 1975, p. 71).

33. *Catalogus codicum astrologorum graecorum* VIII, 4.

34. Ernst Riess, "Nechepsonis et Petosiridis fragmenta magica," *Philologus. Supplement*, vol. 6, 1892, pp. 325-394.

35. Robert Powell, *Hermetic Astrology. Astrology and Reincarnation* (Kinsau, West Germany: Hermetika/AFA, 1987), vol. 1. In the introduction an account of the roots of astrology in the hermetic tradition of ancient Egypt is given, and in Appendix I there is a translation of fragment 14 from Nechepso and Petosiris concerning the hermetic rule ("rule of Hermes") for computing the horoscope of conception.

36. Abraham Sachs, "Babylonian Horoscopes," *Journal of Cuneiform Studies*, vol. 6, 1952, pp. 49-75, traces the emergence of horoscopy among the dating the oldest horoscope known to mankind—excavated from Babylon—to the year 410 BC.

37. Franz Boll, "Hebdomas," *Pauly Real-Encyclopädie der classischen Altertumswissenschaft*, vol. 8, part II, col. 2572. The origin of the planetary week with the hermetic savants of Egypt is substantiated by Dion Cassius (third century AD): "The dedication of the days to the stars called planets originated in Egypt" (*Historiarum* XXXVII, 18).

38. Justin Martyr, Apology I, 67 (quoted from F.H. Colson, *The Week*, Cambridge, England: Cambridge University Press, 1926).

39. *Catalogus codicum astrologorum graecorum* IV, 99

40. Porphyry, *Vita Pythagorae*, 6 (ed. Nauck): "At Babylon Pythagoras associated with the Chaldeans and visited Zaratas by whom he was cleansed of the pollutions of his earlier life."

41. P.F. Gössmann, "Planetarium Babylonicum," *Sumerisches Lexikon* (Rome: Pontifical Institute, 1950), p. 53.

42. Philip Mayerson, Classical Mythology in Literature, *Art, and Music* (Toronto: Xerox Corporation, 1971).

43. Robert Powell, *Hermetic Astrology*, chapter 10 deals with the application of the hermetic rule to determine the moment of conception retrospectively from the moment of birth. The embryonic period according to the hermetic rule is reckoned to be on average ten lunar sidereal periods in length, i.e. $10 \times 27^1/_3 = 273$ days.

44. Hans Lewy, *Chaldaean Oracles and Theurgy* (Cairo: University of Cairo, 1956), p. 90.

45. Ibid., p. 151.

46. "Hymn to King Helios," *The Works of the Emperor Julian* (trsl. W.C. Wright, 3 vols., Cambridge, MA and London: Loeb Classical Library, 1913-1923), vol. 1, pp. 359-361.

47. A. Wilmart, "La collection des 38 homélies latines de St. Jean Chrysostome," *Journal of Theological Studies*, vol. 19, 1918, pp. 305-327.

48. David Pingree, "Astrology," *Dictionary of the History of Ideas*, vol. 1, p. 121.

49. *The Yavanajataka of Sphujidhvaja* (ed. and trsl. D. Pingree, 2 vols., Cambridge, MA and London: Harvard University Press, 1978).

50. Manilius, *Astronomica* II, 922-925 (trsl. G.P. Goold, Cambridge, MA and London: Loeb Classical Library, 1977, p. 155).

51. Ibid., p. 157.

52. See footnote 36.

53. Robert Powell, *Hermetic Astrology*, op. cit., chapter 4.

54. Hans Lewy, p. 45.

55. Ibid., p. 416.

56. Macrobius, *Commentary on the Dream of Scipio* I, xii, 13-14 (trsl. W. Stahl, New York: Scribner's Sons, 1952, p. 136).

The charts, reports and most books listed in *All About Astrology* booklets
are available through:

> Astro Communications Services, Inc.
> 5521 Ruffin Road
> San Diego, CA 92123
> 1-800-888-9983

ISBN 0-935127-02-0

50400

9 780935 127027